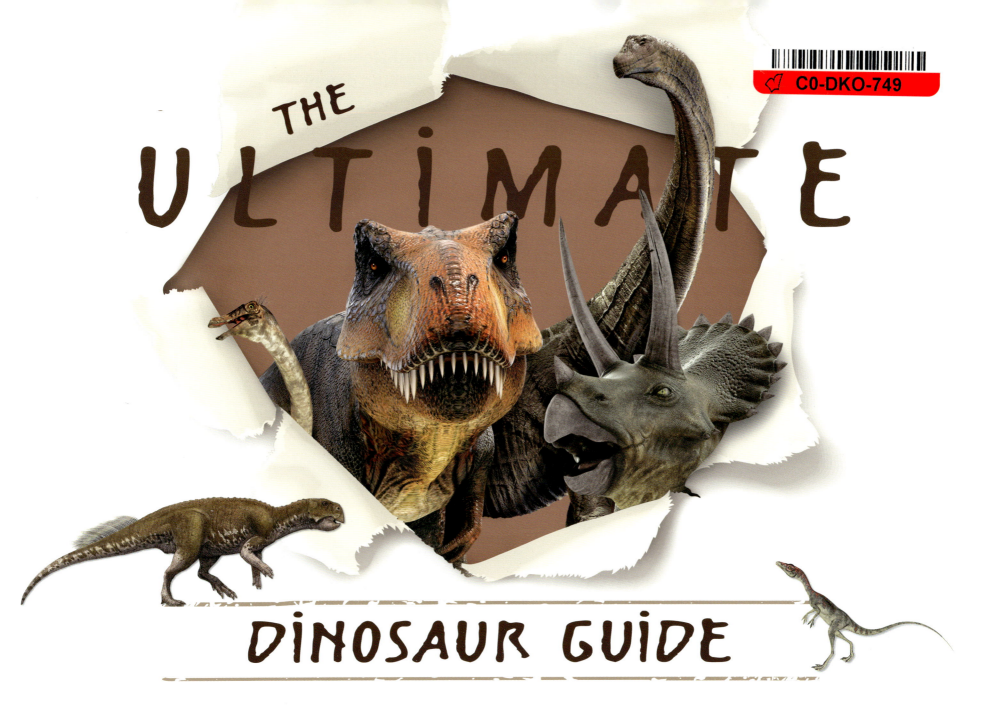

THE ULTIMATE DINOSAUR GUIDE

Written by Heather Hawkins • Designed by Robin Fight

All dinosaurs belong to one of two orders: **Saurischia** or **Ornithischia**. Dinosaurs of the order Saurischia, meaning "lizard-hipped," have pubis bones pointing downward and to the front, like those of lizards. Dinosaurs belonging to the order Ornithischia, which means "bird-hipped," have hip bones shaped like those of birds, with the pubis bone pointing downward and toward the tail.

SAURISCHIA

Saurischian ("lizard-hipped") Hip Bones

Pelvis (Ilium and Ischium bones)

Pubis

The order Saurischia is made up of two groups of dinosaurs: *Sauropods* and **Theropods**. These groups can be divided into several genuses. Some of the most common genuses for each group are shown to the left. Details for these genuses can be found on the pages listed next to each genus name.

SAUROPODS

Alamosaurus | pages 6–7

Apatosaurus | pages 8–9

Brachiosaurus | pages 10–11

Camarasaurus | pages 12–13

Diplodocus | pages 14–15

Plateosaurus | pages 16–17

THEROPODS

Allosaurus | pages 20–21

Compsognathus | pages 22–23

Gallimimus | pages 24–25

Spinosaurus | pages 26–27

Troodon | pages 28–29

Tyrannosaurus rex* | pages 30–31

Velociraptor | pages 32–33

*Unlike the other genuses shown on these pages, *Tyrannosaurus rex* is a distinct species belonging to the genus *Tyrannosaurus*.

ORNITHISCHIA

Ornithischian ("bird-hipped") Hip Bones

Pelvis (Ilium and Ischium bones)

Pubis

 **ANKYLOSAURS**

Ankylosaurus | pages 36–37

Edmontonia | pages 38–39

Euoplocephalus | pages 40–41

Polacanthus | pages 42–43

 **CERATOPSIANS**

Protoceratops | pages 46–47

Psittacosaurus | pages 48–49

Styracosaurus | pages 50–51

Triceratops | pages 52–53

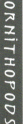 **ORNITHOPODS**

Corythosaurus | pages 56–57

Heterodontosaurus | pages 58–59

Iguanodon | pages 60–61

Maiasaura | pages 62–63

Ornithischian dinosaurs can be divided into five groups: **Ankylosaurs, Ceratopsians, Ornithopods, Pachycephalosaurs,** and **Stegosaurs**. The Ornithischian groups can be divided into several genuses. Some of the most common genuses for each group are shown above and to the right. Details for these genuses can be found on the pages listed next to each genus name.

 PACHYCEPHALOSAURS

Pachycephalosaurus | pages 66–67

Stegoceras | pages 68–69

 **STEGOSAURS**

Kentrosaurus | pages 72–73

Stegosaurus | pages 74–75

Tuojiangosaurus | pages 76–77

DEFINITIONS

Carnivore

an animal that eats meat

Herbivore

an animal that eats plants

Omnivore

an animal that eats meat and plants

Carnosaur

any large theropod dinosaur of the Carnosauria group

Biped(al)

an animal with two legs

Quadruped(al)

an animal with four legs

Paleontologist

a scientist who studies the life and fossils of the past

Bone bed

an area containing a large number of the same kind of dinosaur fossils

SAUROPODS

...inosaurs include Alamos...
...Diplodocus, and ...

...Camaras...
...de

...patosaurus, B...
...saurus was a large, pla...
...herbivorous quadruped t...
...were massive, thick-necked...
...opods, held their bodies...
...pright rather than para...
...the ground.

ALAMOSAURUS

[al–uh–moe–SOR–us]

*A*lamosaurus shared many traits with other Sauropods. It was a large, herbivorous quadruped that had a long neck and tail. *Alamosaurus*, though, was much broader and heavier than other Sauropods such as *Diplodocus* and *Apatosaurus*. *Alamosaurus* is part of a subgroup of Sauropods called titanosaurs, which were massive, thick-necked dinosaurs that, unlike other Sauropods, held their bodies upright rather than parallel to the ground. *Alamosaurus* is the only titanosaur that has been found in North America.

Conceptual illustration of *Alamosaurus* skull

CLASSIFICATION

Order   *Saurischia*

Group   *Sauropod*

LENGTH   *20–24 m (65–78 ft)*

WEIGHT   *45,000 kg (100,000 lb)*

NAME MEANING   *"Alamo Lizard"*

FOUND   *North America (New Mexico, Texas, Utah, Wyoming)*

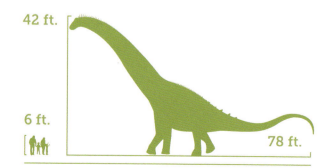

42 ft.

6 ft.

78 ft.

FOSSIL STUDY

Can you believe these gigantic bones are neck vertebrae? Found in Big Bend National Park in Texas, they were the first *Alamosaurus* neck bones ever discovered. A team readied the neck vertebrae for transport and used a helicopter to lift them out of the park one large bone at a time. They were then put on a flatbed truck and taken to the museum for restoration.

Ready for transport

Alamosaurus vertebrae at Big Bend National Park

Though you may suspect it was named after the famous Alamo in San Antonio, Texas, *Alamosaurus* is named for the area in which it was first discovered—the Ojo Alamo Formation in New Mexico. Unfortunately, as is the case with many other large dinosaurs, we do not have any complete adult specimens. In fact, no *Alamosaurus* skull has yet been found, although slender teeth unearthed near *Alamosaurus* specimens are believed to belong to the plant-eating titanosaur.

Scientists suspect that *Alamosaurus* may have been (at least partially) armored. A recently found juvenile specimen suggests the dinosaur may have had bony deposits under the skin that formed plates. Today's crocodiles have a similar type of body armor.

FASCINATING FACT

Alamosaurus may have been the largest land animal to have ever walked the earth!

APATOSAURUS [uh–PAT–uh–SOR–us]

17 ft.

6 ft.

75 ft.

What dinosaur was as long as a semi-truck and had a tail it could crack like a whip? *Apatosaurus*! Believed to be one of the largest land animals ever to walk the earth, *Apatosaurus*, on average, could grow to lengths up to 22.8 m (75 ft). This "gentle giant" may have intimidated smaller dinosaurs with its size, but it would not have harmed them; like other Sauropods, it was an herbivore and only would have attacked other dinosaurs in defense.

With its sturdy, pillar-like legs supporting its incredible bulk, the average *Apatosaurus* could weigh as much as six Asian elephants! To maintain its weight, the massive animal would have needed to eat an immense amount of plant material—about 400 kg (880 lb) per day. In fact, scientists speculate that *Apatosaurus* likely had to eat the entire time it was awake in order to consume enough food. Similar to other Sauropods, it had an extremely long neck and tail. It is believed that the dinosaur held these appendages horizontally, parallel to the ground.

Apatosaurus
tail bones

FASCINATING FACT

The second *Apatosaurus* specimen found was thought to be that of a different dinosaur. Scientists classified this dinosaur under a new genus, which was given the name *Brontosaurus*. This name stuck for many decades until scientists determined that the specimen actually was that of an *Apatosaurus*. Because *Apatosaurus* had been the first name given, it won out over *Brontosaurus* as the official genus name. The name *Apatosaurus*, which means "deceptive lizard," describes a genus of dinosaur with vertebrae that look deceptively like those of a swimming reptile.

CLASSIFICATION

Order *Saurischia*

Group *Sauropod*

LENGTH *21–22.8 m (69–75 ft)*

WEIGHT *27,215 kg (60,000 lb)*

NAME MEANING *"Deceptive Lizard"*

Fossilized *Apatosaurus* bones

FOUND *North America (Colorado, Oklahoma, Utah, New Mexico)*

FOOTPRINT IDENTIFICATION

FOSSIL STUDY

The small head of *Apatosaurus* was filled with peg-shaped teeth. It had a square, stout head with nostrils on the top, rather than the front, of the skull. *Apatosaurus* held its head out horizontally about 4 m (13 ft) off the ground. This would have enabled it to easily graze shrubs and trees of varying heights. *Apatosaurus* ate vegetation whole, also swallowing small stones that would have helped with digestion in the stomach.

Apatosaurus excelsus skull

BRACHIOSAURUS

[BRACK–ee–uh–SOR–us]

42 ft.

6 ft.

72 ft.

CLASSIFICATION

Order _Saurischia_

Group _Sauropod_

LENGTH _22 m (72 ft)_

WEIGHT _58,000 kg (128,000 lb)_

NAME MEANING _"Arm Lizard"_

FOUND _North America (Colorado, Wyoming, Utah, Oklahoma)_

FOOTPRINT IDENTIFICATION

FASCINATING FACT

The first _Brachiosaurus_ fossil was a skull found in Colorado in 1883. The paleontologist who found it identified it as belonging to another dinosaur, not aware that he had discovered a whole new genus! In 1900, a partial _Brachiosaurus_ skeleton was found in Colorado, and the genus was finally given its name. It wasn't until 1998 that the skull found in 1883 was identified as having come from a _Brachiosaurus_. Because _Brachiosaurus_ skulls and bodies were not often found together (a direct result of their heads being loosely attached to their necks), it took more than a hundred years to correctly identify the skull.

Can you name an animal found in Africa with a long neck that eats leaves from the tops of trees? If you said a giraffe, you would be right! But there is an ancient animal that can be described in the same way—the _Brachiosaurus_. Like a giraffe, a _Brachiosaurus'_ front legs were longer than its back legs, causing the dinosaur's body to slope down from its shoulders to its short tail. Its long forearms—which gave *Brachiosaurus* its name (meaning "arm lizard")—allowed it to reach high into the trees to eat leaves, its main source of food.

The *Brachiosaurus* is part of a group of dinosaurs called Sauropods, which is distinguished by the massive size and long necks of its dinosaurs. Sauropods lumbered around on thick legs that were like gigantic tree trunks. *Brachiosaurus* needed sturdy legs to support its heavy body, since a full-grown adult could weigh up to 58,000 kg (128,000 lb)!

Finestra beneath each eye socket

FOSSIL STUDY

The skull of *Brachiosaurus* reveals information about how this gigantic creature lived. Its large nostrils were on top of its head, suggesting that *Brachiosaurus* had a very good sense of smell, perhaps even better than its eyesight. The two openings on either side of the eye socket are called *finestra*. These openings are very large in *Brachiosaurus*, scientists believe, in order to allow its powerful jaw muscles room to move and flex.

Brachiosaurus skull

Large nasal openings on top of skull

CAMARASAURUS

[KAM–uh–ruh–SOR–us]

25 ft.

6 ft.

65 ft.

The most common Sauropod to be found in North America, *Camarasaurus* made its home in the western United States. It is considered to be small for a Sauropod, weighing "only" 20 tons! There are two possible reasons that so many *Camarasaurus* specimens have been found—both due to the dinosaur's smaller size. One possibility is that, because it was smaller, it needed less food and was easily supported by the limited resources in the environment. Another possibility is that, due to its smaller size, its bones were quickly covered by sediment and were, therefore, better preserved.

Camarasaurus' diet largely consisted of conifers, found in abundance in the western United States at the time the dinosaur lived. It roamed in herds, scouring the landscape for the tall tree canopy for which its extended, upright neck and strong, spoon-shaped teeth were suited. Unlike other Sauropods whose weak teeth required softer vegetation found lower to the ground, *Camarasaurus* could subsist on higher roughage alone. It's likely that the animal could break off entire branches with its powerful bite.

Skull of a
Camarasaurus

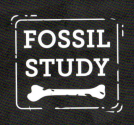

FOSSIL STUDY

The specimen in this picture is the most complete Sauropod specimen ever discovered! It even has intact ear bones! It was so perfect, scientists didn't fully remove it from the stone around it; they only straightened its back legs and tail for display.

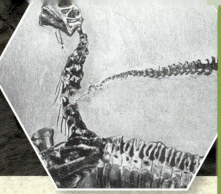

CLASSIFICATION

Order _Saurischia_

Group _Sauropod_

LENGTH _15–23 m (49–75 ft)_

WEIGHT _9,070 kg (40,000 lb)_

NAME MEANING _"Chambered Lizard"_

FOUND _North America (Wyoming, Utah, Colorado, New Mexico, Montana)_

FOOTPRINT IDENTIFICATION

Like many Sauropods, _Camarasaurus_ swallowed its food whole. To aid its digestion, the dinosaur would swallow small stones that helped break down the rough plant material in its stomach. These stones, called gastroliths, have been found in many Sauropod specimens. Once the stones were smooth, they would continue on through the digestive tract. Chickens also use gastroliths to help digest food.

FASCINATING FACT

Many dinosaurs laid their eggs in nests, a fact evidenced by fossilized eggs being found in groups. _Camarasaurus_ eggs, however, have been found arranged in lines. This suggests that the beast did not spend a great amount of time tending to its eggs or young.

DIPLODOCUS

[dih–PLOD–uh–kuss]

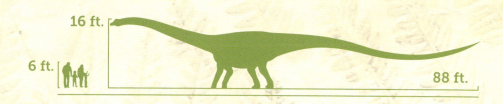

16 ft.

6 ft.

88 ft.

Like all Sauropods, *Diplodocus* was an herbivore that walked on four legs. It had an extremely long tail and neck and a very small head relative to the rest of its body. *Diplodocus* had a double row of 88 bones down the length of its tail! The bones along the bottom of its tail are beam shaped, a unique feature that gave *Diplodocus* its name, which means "double beam."

CLASSIFICATION

Diplodocus tail bones

Order *Saurischia*

Group *Sauropod*

LENGTH *27 m (88 ft)*

WEIGHT *12,000 kg (26,455 lb)*

NAME MEANING *"Double Beam"*

FOUND *North America (Colorado, Wyoming, Utah, New Mexico)*

○○○ FASCINATING FACT

When British King Edward VII saw a picture in Andrew Carnegie's Scottish castle of the newly found *Diplodocus* specimen, the king requested a copy. Carnegie donated a cast to London's Natural History Museum in 1905, which "Dippy," as it came to be nicknamed, has called home ever since.

FOOTPRINT IDENTIFICATION

Scientists believe that *Diplodocus* used its whip-like tail in defense, whipping its tail to make a loud cracking sound (one that could burst the eardrums of its enemies) in warning. Its tail could move at speeds of up to 1,200 km/hr (745 mph)! Even its most fearsome predator, the mighty *Allosaurus*, likely was not a match for *Diplodocus'* deadly tail.

Although all Sauropods are extremely long, *Diplodocus* is the longest of all the dinosaurs for which we have complete skeletons. In fact, *Diplodocus* was longer than a tennis court and one of the longest land animals that has ever lived!

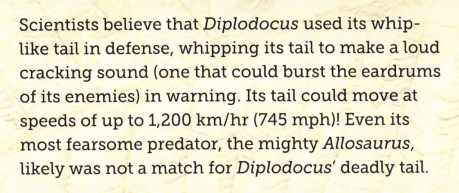

FOSSIL STUDY

Diplodocus teeth

Different from the other dinosaurs in its group, *Diplodocus* had teeth that angled outward. These teeth, which look like pegs bunched up at the front of its mouth, were used to strip leaves from branches. Because *Diplodocus* needed so much vegetation to feed itself, its teeth became worn down and often fell out. It was able to grow new teeth, though!

Roaming around on its stocky legs, *Diplodocus* looked for enough plant matter to fill its belly. Its diet consisting of conifer trees, ferns, cycads, and mosses, the dinosaur would have used its peg-like teeth to strip off leaves and swallow them whole.

PLATEOSAURUS

[PLAT–ee–oh–SOR–us]

6 ft.   10 ft.   26 ft.

CLASSIFICATION

Order _Saurischia_

Group _Sauropod_

LENGTH _7.9 m (26 ft)_

WEIGHT _1,815 kg (4,000 lb)_

NAME MEANING _"Flat Lizard"_

FOUND _Central and Northern Europe (France, Germany, Switzerland, Greenland)_

FOOTPRINT IDENTIFICATION

Plateosaurus footprints

Plateosaurus holds the honor of being one of the dinosaurs about which scientists know the most. There are more than 100 different specimens that have been found, many of them nearly complete, giving paleontologists a good understanding of the animal's habits and lifestyle. One of the first dinosaur fossils ever found was a _Plateosaurus_ specimen—the first unearthed in Germany in 1834! Only three years later, in 1837, _Plateosaurus_ was given its name, which means "flat lizard." This was four years before the word "dinosaur" was even coined!

FOSSIL STUDY

Plateosaurus had an unusually large, sickle-like claw on each thumb. Scientists speculate that this claw could have been a ready weapon for defense or a useful tool, perhaps for sawing off tough vegetation. Can you think of some other ways a large thumb claw could be used?

Plateosaurus claws

FASCINATING FACT °°°

Plateosaurus would not have been able to run in a typical way; it always would have needed to have one foot on the ground. Despite this trait, *Plateosaurus* could increase its stride in order to reach speeds of up to 64 km/hr (40 mph)!

Plateosaurus is believed to have been a herd animal, which means it traveled in large groups. Moving in herds gave *Plateosaurus* a better chance of survival because an entire herd was more likely to spot predators than a solitary dinosaur was. *Plateosaurus* could stand on its two back legs to access high branches or rest on all fours to graze on lower vegetation. Its teeth were broad and short—perfect for crushing vegetation—while its long, straight tail counterbalanced the length and weight of its neck and skull.

Most of the *Plateosaurus* specimens we have today come from three similar areas in Europe, each called a *bone bed* due to the abundance of dinosaur specimens found there. The skeletons from these bone beds, most of them intact and complete, are almost all *Plateosaurus* adults. Many of these skeletons were found with their feet below their bodies, as if the animals had died while standing. Scientists believe these heavy beasts would get trapped in the thick mud of water holes, sinking deeper as they struggled to free themselves and ultimately dying in an upright position. Theropod teeth also have been found in these bone beds, likely lost when Theropods scavenged the *Plateosaurus* remains.

SAUROPODS GROUP GUIDE

CHARACTERISTICS

Sauropods (meaning "lizard-footed") were massive, herbivorous beasts, the majority of which were quadrupeds (walking on all fours). Although a few Sauropod genuses were able to hold their long necks upright, most were structured with their necks and bodies parallel to the ground. Thick and sturdy legs supported their enormous bodies.

FOSSIL SITES

Fossils of Sauropod dinosaurs have been found on all continents, even Antarctica.

SIZE COMPARISON

42 ft.

6 ft.

Plateosaurus

Diplodocus

Apatosaurus

Camarasaurus

Brachiosaurus

Alamosaurus

THEROPODS

...pod dinosaurs include *Allosaurus*... *Spinosaurus*, *Tre*... ...Therop... ...dinosaur... ...nus *S*... ...*Veloci*...

ALLOSAURUS

[AL–oh–SOR–us]

15 ft.

6 ft.

39 ft.

CLASSIFICATION

Order *Saurischia*

Group *Theropod*

LENGTH *8.5 m–12 m
(28–39 ft)*

WEIGHT *2,300 kg (5,070 lb)*

NAME MEANING *"Different Lizard"*

FOUND *Mostly North America; recent
finds have also occurred in Europe*

FOOTPRINT IDENTIFICATION

At around 11 m (36 ft) long from the tip of its thick tail to the front of its fearsome jaws, *Allosaurus* was one of the largest carnosaurs (meat-eating dinosaurs). *Allosaurus*, whose name means "different lizard," had two powerful back legs that propelled it forward at running speeds of up to 55 km/hr (34 mph) when chasing prey. That's similar to the speed of a grizzly bear! Its short arms had three curved, pointed claws that were 15 cm (6 in) long and were used as hooks to grab and tear meat.

FASCINATING FACT

Allosaurus is the official state fossil of Utah.

FOSSIL STUDY

Scientists believe that *Allosaurus* and *Stegosaurus* were bitter enemies. Broken teeth from *Allosaurus* are often found scattered around the bones of *Stegosaurus* at archaeological digs, likely lost when *Allosaurus* was feeding on *Stegosaurus*. In some cases, injury markings in portions of *Stegosaurus* bones even can be matched to *Allosaurus* jaws in size and shape. *Allosaurus* bones have also been found with injuries that match the tail spikes of *Stegosaurus*.

Allosaurus teeth

Fossilized *Allosaurus* jaw

Allosaurus skull

These massive predators were at the top of their food chain, meaning that they had no predators themselves. *Allosaurus* hunted large Sauropods, smaller plant eaters, and possibly even other carnosaurs. *Allosaurus* was considerably smaller than its Sauropod prey, which leads scientists to believe it may have hunted in packs.

A great number of *Allosaurus* fossils have been found in sites in America, namely Utah, Wyoming, and Colorado, making *Allosaurus* one of the most common dinosaurs to have lived in North America. The most famous *Allosaurus*, known as "Big Al," is a well-preserved, nearly complete specimen found in Wyoming in 1991. Examining the bones of Big Al seems to indicate that he had a rough life. He had 18 healed bone fractures and another fresh fracture that had gotten infected and likely caused his death. Big Al even left what scientists believe are the remains of his last meals—a tooth from a lungfish, a piece of a hip bone from a small herbivorous dinosaur, and other bone fragments.

COMPSOGNATHUS

[komp–sog–NAY–thus]

6 ft.

1.5 ft.

3 ft.

CLASSIFICATION

Order *Saurischia*

Group *Theropod*

LENGTH *0.6–1.2 m (2–4 ft)*

WEIGHT *1–4 kg (2–10 lb)*

NAME MEANING *"Elegant Jaw"*

FOUND *Western Europe (Germany, France)*

FOOTPRINT IDENTIFICATION

The first nearly complete dinosaur ever found, *Compsognathus* holds a unique place in the world of dinosaurs. For many years, it was the smallest known dinosaur, being only as big as a chicken! Even as small as it was, it was the largest dinosaur in its habitat! The other ancient creatures in the region were not technically dinosaurs. There was *Archaeopteryx*, classified as an ancient bird, and also lots of ancient fish and reptiles.

Only two specimens of *Compsognathus* have been found, one in modern-day Germany and one in France, but they are nearly complete skeletons. These well-preserved fossils have given scientists a full picture of the lifestyle and eating habits of *Compsognathus*. Classified as a Theropod, the Compy, as it is sometimes called, had similar characteristics (including size and body shape) to other bird-like dinosaurs, leading scientists to believe it may have had feathers.

You might be asking yourself, "How could such a tiny creature be classified in the same group as fierce *Tyrannosaurus rex* and gigantic *Allosaurus*?"

The answer is simple. Like all other Theropods, *Compsognathus* had hollow bones, three-fingered limbs, short forearms, and was bipedal. It was also carnivorous, a trait typical to most Theropods. *Compsognathus'* diet consisted of insects and small reptiles. In order to be able to catch such agile prey, *Compsognathus* necessarily had to have been extremely quick footed with keen eyesight. Its long tail aided its balance and helped it make quick turns as it darted after prey.

FASCINATING FACT ● ● ●

Compsognathus was originally thought to be a lizard, one of the "most curious forms among the lizards," according to the first paleontologist who examined the specimen.

FOSSIL STUDY

Both of the *Compsognathus* specimens found had the remains of lizards (one of *Compsognathus'* main food sources) in their bellies. The photo to the right is of the German specimen found in 1859, which was the most complete dinosaur fossil that had ever been found at the time.

GALLIMIMUS [gal–i–MY–mus]

A long time ago, the arid Gobi Desert in modern-day Mongolia was a wet lowland habitat with abundant rivers and lakes. In the midst of this humid paradise lived a creature that jaunted around on its two slim back legs to snatch up plants or other animals' eggs with its short forearms and sharp claws. This ancient animal was a type of Theropod, yet very different from its terrifying *T. rex* cousin. In fact, the bird-like *Gallimimus* was much more likely to be prey than a predator.

Gallimimus is often compared to an ostrich, and one can see the resemblance when examining its long, slender legs with three-clawed, chicken-like toes. Like an ostrich, it was a fast runner, perhaps reaching speeds of 55 km/h (34 mph), fast enough to outrun its predators.

Paleontologists have debated whether *Gallimimus* was a carnivore, herbivore, or omnivore. Because the environment in which it lived contained lots of water, recent speculation is that *Gallimimus* used its toothless bill like a duck, scooping up mouthfuls of water and straining out plants, mollusks, and other water organisms to eat.

11 ft.

6 ft.

22 ft.

FOSSIL STUDY

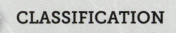

The pelvic bone, or ilium, of a *Gallimimus*

Despite *Gallimimus* diverging from the stereotypical idea of a ferocious, meat-eating dinosaur, it is indeed classified with other Theropods such as *Tyrannosaurus rex* and *Allosaurus*. Theropods and another group of slow-moving, large, plant-eating dinosaurs called Sauropods are both classed into the order known as Saurischia. This order, which means "lizard-hipped," includes all dinosaurs whose hip bones were shaped like those of lizards. The ilium is the pelvic bone of the dinosaur.

CLASSIFICATION

Order *Saurischia*

Group *Theropod*

LENGTH *3.4–8 m (11–26 ft)*

WEIGHT *118 kg (260 lb)*

NAME MEANING *"Chicken Mimic"*

FOUND *Asia (Mongolia)*

FOOTPRINT IDENTIFICATION

FASCINATING FACT

These two *Gallimimus* skeletons were illegally smuggled out of Mongolia by an American private collector using forged customs papers. They were soon discovered and seized by US authorities along with 16 other dinosaur specimens and given back to the Mongolian government in 2014. They are now back in their home country and on display at the Central Museum of Mongolian Dinosaurs.

SPINOSAURUS

[SPY-no-SOR-us]

22 ft.

6 ft.

54 ft.

Like all other Theropods, *Spinosaurus* was a bipedal (walking on two legs) carnivore, although its longer arms suggest that it may have occasionally walked on all four limbs. *Spinosaurus* lived in the coastal mangroves of Egypt and Northern Africa and spent considerable time in the water. It had long spines protruding from its back that were covered with skin, forming a fin-like sail. The sail alone would have been more than 2 meters (6.5 feet) tall—taller than most adults! *Spinosaurus* is believed to be the largest carnivorous dinosaur, even bigger than *Tyrannosaurus rex*!

CLASSIFICATION

Spinosaurus teeth

Order *Saurischia*

Group *Theropod*

LENGTH *14–18 m (46–59 ft)*

WEIGHT *12,000–20,000 kg (26,000–44,000 lb)*

NAME MEANING *"Spined Lizard"*

FOUND *Northern Africa*

FASCINATING FACT

The remains of the first discovered *Spinosaurus* skeleton were housed in a museum in Munich, Germany, until they were destroyed by Allied bombing during World War II.

FOOTPRINT IDENTIFICATION

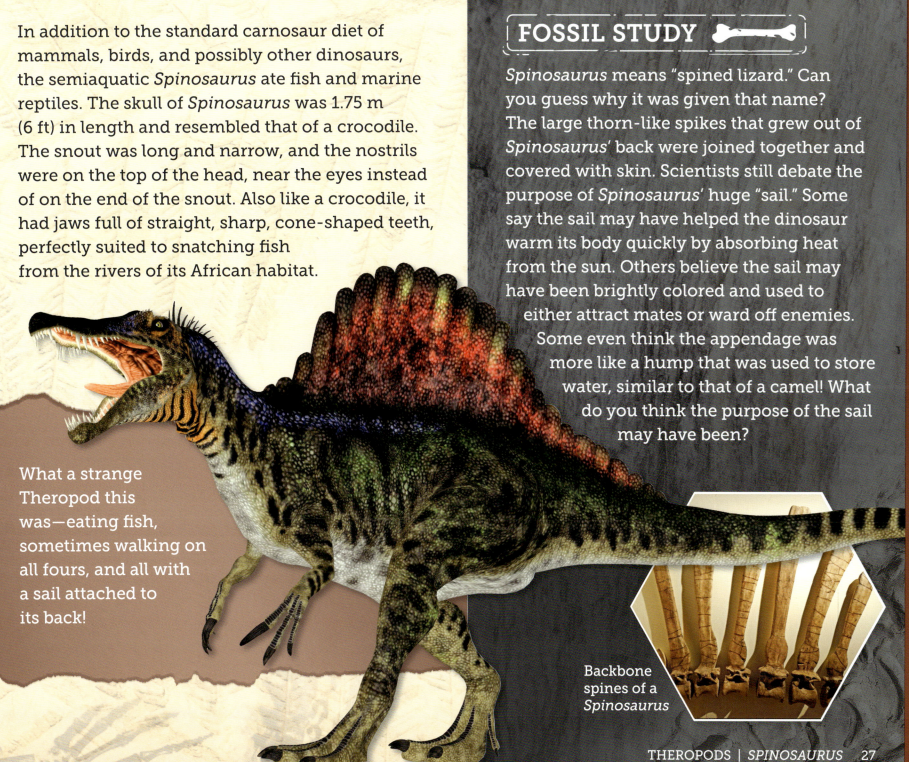

In addition to the standard carnosaur diet of mammals, birds, and possibly other dinosaurs, the semiaquatic *Spinosaurus* ate fish and marine reptiles. The skull of *Spinosaurus* was 1.75 m (6 ft) in length and resembled that of a crocodile. The snout was long and narrow, and the nostrils were on the top of the head, near the eyes instead of on the end of the snout. Also like a crocodile, it had jaws full of straight, sharp, cone-shaped teeth, perfectly suited to snatching fish from the rivers of its African habitat.

What a strange Theropod this was—eating fish, sometimes walking on all fours, and all with a sail attached to its back!

FOSSIL STUDY

Spinosaurus means "spined lizard." Can you guess why it was given that name? The large thorn-like spikes that grew out of *Spinosaurus*' back were joined together and covered with skin. Scientists still debate the purpose of *Spinosaurus*' huge "sail." Some say the sail may have helped the dinosaur warm its body quickly by absorbing heat from the sun. Others believe the sail may have been brightly colored and used to either attract mates or ward off enemies. Some even think the appendage was more like a hump that was used to store water, similar to that of a camel! What do you think the purpose of the sail may have been?

Backbone spines of a *Spinosaurus*

TROODON

[TRO-uh-don]

6 ft.

5 ft.

6.5 ft.

A small, light, bird-like dinosaur, *Troodon* stood less than 1 m (3 ft) tall at its hips. It was a swift-moving, agile dinosaur that easily would have been able to outrun most of the other creatures of its time. *Troodon* had long arms that folded back like a bird's, with thin, sharp claws and partially opposable thumbs that would allow it to grasp food. After chasing down its prey with its powerful hind legs, *Troodon* could thrust forward its sickle-like claw on its second toe to wound its prey. Its prominent eyes enabled it to see well in low light, and its speed and telescopic vision would have made it an effective night hunter. *Troodon* fed on small invertebrates, mammals, and reptiles with its mouth full of small, serrated teeth. Because teeth of this size and shape commonly are associated with plant eaters, scientists believe that *Troodon* may have been omnivorous, an unusual trait for a dinosaur belonging to the mostly carnivorous group of Theropods.

Troodon skull

FASCINATING FACT ∘∘∘

Troodon had one of the largest brains in comparison to its body, up to six times the size of other dinosaurs of its time. Scientists think it may have been the most intelligent of all the dinosaurs! Unfortunately for *Troodon*, it still would have been no smarter than a chicken!

FOSSIL STUDY

We know many details about the nesting and parenting habits of *Troodon*. Nesting grounds were discovered in Montana, revealing full nests of 16–24 eggs arranged in a circle. Scientists believe that the female *Troodon* would lay two to four teardrop-shaped eggs a day for about a week, setting them upright and half buried in the dirt. There is evidence that after the female laid the eggs, the male would sit on and incubate the eggs.

Fossilized *Troodon* eggs

Troodon tooth

CLASSIFICATION

Order *Saurischia*

Group *Theropod*

LENGTH *2 m (6.5 ft)*

WEIGHT *40 kg (90 lb)*

NAME MEANING

"Wounding Tooth"

FOUND *North America (Canada, Alaska, Montana, and other western US states)*

FOOTPRINT IDENTIFICATION

When the first *Troodon* fossil was discovered in Montana in 1855, a surveyor who unearthed the fossil sent a single, serrated tooth from the specimen to famous American naturalist Joseph Leidy. Leidy thought a new genus of lizard had been discovered, and he named the "lizard" *Troödon*—meaning "wounding tooth." The name was eventually simplified to *Troodon*. In the early 1930s, dinosaur hunters began finding other *Troodon* fossils, including fragments from a hand, a foot, and some tail vertebrae. It wasn't until 1983, when a paleontologist discovered a jawbone with the same type of serrated teeth, that scientists were able to piece together what *Troodon* may have looked like.

TYRANNOSAURUS REX

[tie–RAN–uh–SOR–us REKS]

20 ft.

6 ft.

42 ft.

CLASSIFICATION

Order _Saurischia_

Group _Theropod_

LENGTH _12.8 m (42 ft)_

WEIGHT _6,350 kg_
_(14,000 lb)_

NAME MEANING _"Tyrant Lizard King"_

FOUND _North America and Mongolia_

T. rex skull

FOOTPRINT IDENTIFICATION

The famous _Tyrannosaurus rex_, the "tyrant lizard king," was a massive, meat-eating dinosaur that mercilessly ruled the land during its time. With its powerful hind legs, relatively tiny arms, and huge head and jaws, it is one of the most recognizable dinosaurs. _Tyrannosaurus rex_ (also known as _T. rex_) was the largest land carnivore of its time and the third-largest of all time (only bested by _Spinosaurus_ and the rightly named _Giganotosaurus_). When it stood at full height, it was nearly 6 meters (20 feet) tall, as tall as a two-story house!

FASCINATING FACT

One of the largest and best-preserved _Tyrannosaurus rex_ specimens ever found, nicknamed "Sue," was seized by the US Government when it was discovered that Sue had been found on Indian Territory held in trust by the US Department of the Interior. The specimen was eventually sold to the Chicago Field Museum for a record $8,362,500!

FOSSIL STUDY

When *Tyrannosaurus rex* walked, it left footprints that were nearly 1 m (3 ft) long! Scientists have identified some *T. rex* footprints by the presence of an imprint of a fourth toe (called a hallux). A set of prints found in New Mexico that were previously attributed to a plant-eating dinosaur were later identified as those of a *T. rex* after a hallux imprint was recognized.

However, the most awe-inspiring trait was not the height of *Tyrannosaurus rex*, but the size of its fearsome head and jaws. The head of a *T. rex* could span 1.5 m (5 ft) in length. Its mighty jaws held 60 razor-sharp teeth, some measuring up to 23 cm (9 in) in length. Scientists believe that the jaws of an adult *Tyrannosaurus rex* were so powerful they could pierce the toughest hides and even rip off the horns of a *Triceratops*!

Because *Tyrannosaurus rex* grew so large during its 30-year lifespan, it's likely that its hunting habits changed to accommodate its increasing size. Smaller and more agile in its youth, the juvenile *T. rex* easily could have chased after small, speedy prey. As it aged, however, it grew so massive that it would have had to move more slowly. In addition, its short neck became more muscular to support its heavy head, its jaw muscles strengthened, and its stout tail grew even thicker to help balance its enormous body. Although the juvenile *T. rex* was faster, it did not yet have the jaw strength nor the size to take down stronger and larger prey. Conversely, the fully-grown *T. rex* preyed on slower, larger dinosaurs. It could reach speeds of up to 19 km/hr (12 mph) while making its attack.

T. rex footprint

VELOCIRAPTOR [vel–OS–i–rap–tor]

Velociraptor, whose name means "swift thief," is part of the family of dinosaur raptors. Raptors were small- to medium-sized, feathered dinosaurs that had a large, sickle-shaped talon on the second toe of each foot. *Velociraptor* had a long skull and a flat snout. About the size of a turkey and weighing up to 45 kg (100 lb), *Velociraptors* resembled modern-day predatory birds. They were also part of the Theropod group, meaning they, too, were carnivorous dinosaurs with hollow bones and three-toed feet. *Velociraptors* lived only in Asia, in a dry and arid desert climate.

Despite its small size, *Velociraptor* was a fierce predator. It used its enlarged, deadly claws to hold down and tear into its prey. Paleontologists have found a complete *Velociraptor* fossil locked into that of a *Protoceratops*, a horned dinosaur that was bigger and stouter than the bird-like *Velociraptor*. It is believed both dinosaurs were killed during their vicious fight when they were overtaken by a sudden sandstorm, which preserved their fossils well.

6 ft.

3 ft.

6 ft.

FASCINATING FACT •••

Velociraptors had strong, rigid tails that helped them balance while running and attacking. At full extension, the tail of a *Velociraptor* was about 1 m (3 ft) long!

It's likely that *Velociraptors* even attacked each other. A *Velociraptor* skull has been found with two rows of wound indentations that align exactly with *Velociraptor* teeth. Scientists believe the skull wounds are proof that *Velociraptors* fought each other, even to the death.

Scientists also insist that *Velociraptor* must have been warm-blooded. Two features support this: 1) *Velociraptor* was feathered, and 2) it expended a LOT of energy running and hunting. Animals with feathers tend to be warm-blooded because the feathers act as insulation against the cold. Warm-blooded animals also need to regulate their temperatures by frequently hunting and eating.

CLASSIFICATION

Order   *Saurischia*

Group   *Theropod*

LENGTH   *1.8 m (6 ft)*

WEIGHT   *Up to 45 kg (100 lb)*

NAME MEANING   *"Swift Thief"*

FOUND   *Asia (Mongolia and China)*

Velociraptor claw

FOOTPRINT IDENTIFICATION

FOSSIL STUDY

Velociraptor fossil

Did you know that *Velociraptors* had feathers but couldn't fly? In 2007, a well-preserved *Velociraptor* was uncovered in Mongolia. Upon close examination of the ulna (one of the lower arm bones), scientists discovered quill knobs (bumps on the bones where feathers were attached). Presence of quill knobs is evidence that *Velociraptor* had feathers. Quill knobs also can be seen on a modern-day turkey bone.

THEROPODS GROUP GUIDE

CHARACTERISTICS

Dinosaurs belonging to the Theropod (meaning "beast-footed") group had hollow, thin-walled bones and sharp, blade-like teeth and claws. Theropods were bipedal (walked on two legs) and had three main weight-bearing toes on each foot.

FOSSIL SITES

Fossils of Theropod dinosaurs have been found on all continents.

SIZE COMPARISON

22 ft.

6 ft.

Compsognathus

Velociraptor

Troodon

Gallimimus

Allosaurus

Tyrannosaurus rex

Spinosaurus

ANKYLOSAURS

ANKYLOSAURUS

[AN–kil–uh–SOR–us]

6 ft. 8 ft. 20 ft.

Many plant-eating dinosaurs were at a disadvantage when pitted against their meat-eating predators, but *Ankylosaurus* was an exception. This armored dinosaur, with its heavy defenses and its built-in weapons, is often compared to a tank ready for battle. It was well protected against all its enemies—including the fierce *Tyrannosaurus rex*.

Even though *Ankylosaurus* averaged a lengthy 6 m (20 ft), it sat low to the ground, only rising to about 1.7 m (5.6 ft) tall at the hips. This kept its center of gravity low, which meant that it could not be flipped over easily by larger predators. *Ankylosaurus* had a heavily rounded back covered in oval, bony plates called scutes. These thick plates protected *Ankylosaurus* and were virtually impenetrable to hungry enemies.

CLASSIFICATION

Order *Ornithischia*

Group *Ankylosaur*

LENGTH *6 m (20 ft)*

WEIGHT *3,600 kg (8,000 lb)*

NAME MEANING *"Fused Lizard"*

FOUND *North America (Canada, Montana, Wyoming)*

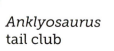

FOOTPRINT IDENTIFICATION

Anklyosaurus tail club

Protruding out of *Ankylosaurus'* body all the way around its sides and back and protecting its legs were long spikes. Its wide head was sheathed with bony armor and had two sets of horns pointing backward. Even its eyes were protected by armor-plated eyelids!

But *Ankylosaurus* was not only a creature armed in defense; it also had one powerful offensive weapon. Its tail was 3 m (10 ft) long and ended in a large, club-like knob. It likely used its tail to swing at the legs of taller predators or in defense against other Ankylosaurs. With all of this heavy-duty protection, it's no wonder that this beast is often compared to a tank ready to battle its enemies!

FASCINATING FACT ∘∘∘

Ankylosaurus only walked about 9.7 km/h (6 mph), which is just barely faster than a turtle at its top speed!

FOSSIL STUDY 🦴

The skull of *Ankylosaurus* was broad and triangular in shape, with a bony covering serving as a kind of helmet. It had a beak filled with small, leaf-shaped teeth that were used for stripping off the leaves of the plants that made up its diet. Recent evidence suggests that *Ankylosaurus* had three-dimensional sight and a very good sense of smell.

Illustration of *Ankylosaurus* tooth

EDMONTONIA

[ed–mon–TOH–nee–uh]

6 ft.   6 ft.

22 ft.

Edmontonia, an Ankylosaur found in the Edmonton Formation in Alberta, Canada, was an armored dinosaur with no clubbed tail. That doesn't mean that it was without protection, though! From its head to its tail, it was covered in ridged, armored plates. It had spikes along its sides, with four large ones positioned behind its shoulders and pointing forward. The plates covering the head, back, and tail were small and oval with ridges, while three rows of larger, ridged plates that were fused together covered its neck and shoulders.

A short neck, short legs, and low-slung body kept Edmontonia close to the ground, eating low-growing vegetation. It had, like other Ankylosaurs, a horned beak and teeth tucked back into its cheeks. These small, ridged teeth would have been perfect for this herbivore's plant diet of ferns and cycads.

Rings of petrified wood from trees native to Edmontonia's environment lend evidence to the theory that the animal's habitat underwent drastic changes between dry and wet seasons. Some Ankylosaur specimens appear to have died during periods of drought, their bones later covered in sediment by the floodwaters of the following wet seasons. This created a perfect environment of preservation, allowing paleontologists to find Edmontonia specimens with the spikes still attached to the bodies.

FOSSIL STUDY

Edmontonia skull

Edmontonia had four shoulder spikes. They may look dangerous, but they probably did not provide much defense against larger predators. Scientists speculate that these spikes may have been similar to deer antlers, likely used when an *Edmontonia* male engaged in battle with another male of its kind. The spikes also could have been used to intimidate smaller creatures or, because they pointed forward, to clear a way through groups of predators when running.

Edmontonia shoulder spikes

CLASSIFICATION

Order Ornithischia

Group Ankylosaur

LENGTH 6.6 m (22 ft)

WEIGHT 3,000 kg (6,600 lb)

NAME MEANING "From Edmonton"

FOUND Canada; Texas

FOOTPRINT IDENTIFICATION

FASCINATING FACT

Edmontonia laid eggs at the perfect time for its young to hatch during the wet seasons, where fresh, green vegetation would be ready and waiting for the hungry baby dinosaurs!

EUOPLOCEPHALUS

[YOU-oh-ploh-SEF-uh-luss]

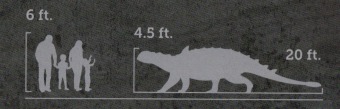

6 ft.

4.5 ft.

20 ft.

CLASSIFICATION

Order _Ornithischia_

Group _Ankylosaur_

LENGTH _6 m (20 ft)_

WEIGHT _1,814 kg (4,000 lb)_

NAME MEANING _"Well-armored Head"_

FOUND _North America (Montana;_
Alberta, Canada); China

○ ○ ○

FASCINATING FACT

Among the numerous _Euoplocephalus_ specimens discovered are fifteen skulls and multiple nearly complete skeletons!

FOOTPRINT IDENTIFICATION

FOSSIL STUDY 🦴

Euoplocephalus tail club

Euoplocephalus, like many Ankylosaurs, had a clubbed tail that it used in defense. Bony tendons supported the club, which was hard like a hammer. Strong muscles toward the end of the tail enabled *Euoplocephalus* to whip its tail around with deadly force.

As is obvious from the meaning of its name, "well-armored head," *Euoplocephalus* was a member of the armored dinosaurs, or Ankylosaurs. It had bony plates and spikes covering its back with plates fused together in a half-ring covering the back of its neck. This quadruped had a broad body, which served two purposes: 1) It prevented predators from easily turning *Euplocephalus* over onto its back to get to its soft belly, and 2) It gave the plant eater a massive space for digestion and for, scientists speculate, a complex fermentation process in which food sat in chambers that would break down the plant matter.

Euoplocephalus had a horny beak with small teeth tucked into its cheeks. Unlike many other Ankylosaurs that had narrow beaks, it had a broad mouth that was wider than it was long. Therefore, *Euoplocephalus* would not have been able to be selective about which plants it ate, chopping off broad clumps of low vegetation. It was not a picky eater!

Euoplocephalus is one of the most important Ankylosaurs due to the large number of fossils that have been discovered. In fact, it is the most well-represented armored dinosaur in North America! Because these specimens have been found one at a time, rather than grouped together, scientists believe that *Euoplocephalus* was solitary and did not roam in herds.

Fossil of armored, plated skin of a *Euoplocephalus*

POLACANTHUS

[pol–uh–CAN–thus]

*P*olacanthus was an armored dinosaur that walked close to the ground on four short legs. It had a huge, shield-like covering, formed from a single sheet of fused bony plates that protected its hips. It also had spikes covering much of its body, which is how it got its name (meaning "many thorns"). All of its armor would make any predator think twice about trying to attack it!

Like many Ankylosaurs, *Polacanthus* had front legs that were shorter than its hind legs. Because of this, the animal could not lift its head more than 1.5 meters (5 feet) off the ground! *Polacanthus* was a plant eater with a diet consisting of low-growing tubers, roots, and fruit. It lived in a warm climate that provided lush vegetation during the rainy parts of the year. During the hot, dry summer months, however, it needed to stay close to ponds and creeks in order to find enough plants to eat.

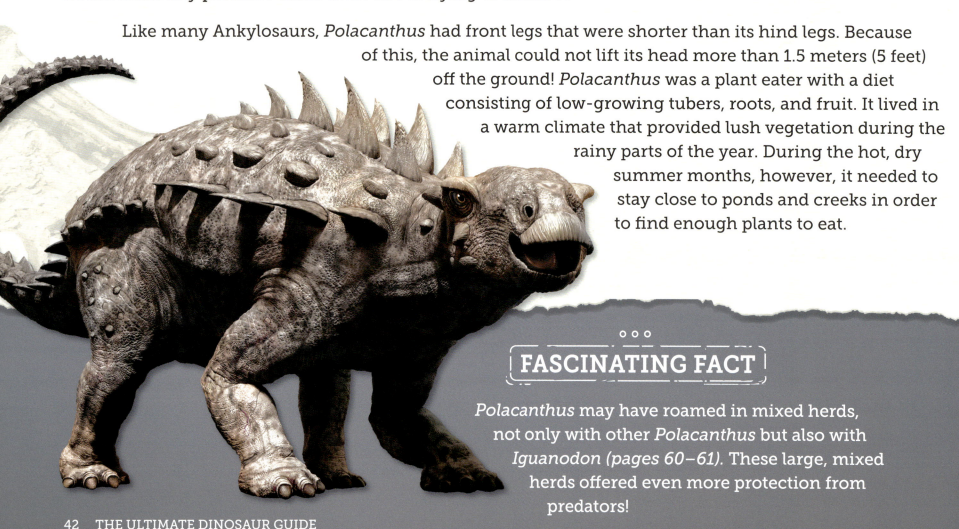

FASCINATING FACT

Polacanthus may have roamed in mixed herds, not only with other *Polacanthus* but also with *Iguanodon (pages 60–61)*. These large, mixed herds offered even more protection from predators!

FOSSIL STUDY

The skin of *Polacanthus* was embedded with osteoderms, which are bony deposits often present in the skin of reptiles and amphibians. These osteoderms formed plates that provided good protection against the bites of predators. Osteoderms also are found on the large, fused bony plate covering the hips of *Polacanthus*, called a "pelvic shield."

Illustration of armored *Polacanthus* pelvis

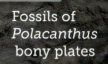

Fossils of *Polacanthus* bony plates

CLASSIFICATION

Order *Ornithischia*

Group *Ankylosaur*

LENGTH *5 m (17 ft)*

WEIGHT *2,000 kg (4,400 lb)*

NAME MEANING *"Many Thorns"*

FOUND *Isle of Wight, England*

FOOTPRINT IDENTIFICATION

Polacanthus was first discovered on the Isle of Wight in 1865 by paleontologist and Reverend William Fox. Specimens of *Polacanthus* have only been found on the Isle of Wight and in southern England, so fossil remains are quite scarce. The most common *Polacanthus* fossils found have been parts of its armor, due to the armor's strong, bony composition.

Fossilized skin impression of a *Polacanthus*

ANKYLOSAURS GROUP GUIDE

CHARACTERISTICS

Ankylosaurs were bulky, quadrupedal plant eaters with fully-armored backs formed from fused bony plates. Most Ankylosaurs had a club at the end of the tail, likely used as a defensive weapon.

FOSSIL SITES

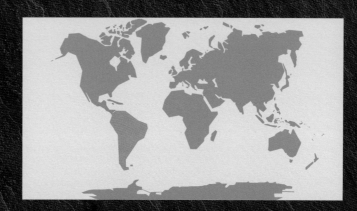

Ankylosaur dinosaur fossils have been found in North America, Europe, Antarctica, Africa, Australia, and Asia.

SIZE COMPARISON

6 ft.

8 ft.

Polacanthus

Euoplocephalus

Edmontonia

Ankylosaurus

CERATOPSIANS

...atopsian dinosaurs include Protocera...
...ttacosaurus, S... ...urus, Styracos...
...nosaurs by the... ...ed parro...
...he m... ...this "...
...ck f...

...face ...had...
...its fac...

...inctive... ...label...
...re distinguished... ...rot dinosaurs by...
...d parrot-like bea... ...most famous...
...distinctive neck frill which...

PROTOCERATOPS

[pro-toh-SAIR-uh-tops]

6 ft. / 3 ft. / 6 ft.

Protoceratops is part of a family of horn-faced dinosaurs, but look at its picture. Do you see a horn on its face? No! Even though it doesn't have a horn, it is still considered part of the Ceratopsian (horn-faced) family. *Protoceratops* does have two distinctive features on its face, however. One is the parrot-like beak that all Ceratopsians have. The other is a bony bump on its nose. Because this bump was larger in older males, scientists believe that it may have been used to headbutt other *Protoceratops* males in contests of strength.

A feature that *Protoceratops* shared with its Ceratopsian family members was the large, bony collar, called a frill, that stood over its neck and upper back. This protective shield grew out of its skull and protected it from attack. The size of *Protoceratops'* frill and of *Protoceratops* in general was much smaller than other Ceratopsians, making the sheep-sized dinosaur one of the smallest of its family.

FASCINATING FACT •••

Protoceratops is one of the few dinosaurs whose males were larger than the females!

CLASSIFICATION

Order *Ornithischia*

Group *Ceratopsian*

LENGTH *1.8 m (6 ft)*

WEIGHT *180 kg (400 lb)*

NAME MEANING *"First Horned Face"*

FOUND *Mongolia, China*

FOOTPRINT IDENTIFICATION

The stiff beak and extremely strong jaws of *Protoceratops* gave this herbivore the tools it needed to successfully rip, tear, and chew the tough plant matter common to its dry, desert habitat. The oversized beak, the large bump on its nose, and its protruding frill made this dinosaur's head look way too big for its body. This feature may have helped ward off would-be attackers by making *Protoceratops* look much more fierce than it actually was. Another means of protection was that *Protoceratops* likely traveled in herds, perhaps numbering in the thousands!

Juvenile skull (top) and adult skull (right)

FOSSIL STUDY

Young *Protoceratops* had no frill, but the bony projection developed and grew larger as the dinosaur aged. In these pictures you see differences between the skulls of a juvenile and an adult *Protoceratops.*

PSITTACOSAURUS

[SIT–uh–coe–SOR–us]

6 ft.

2 ft.

6.5 ft.

CLASSIFICATION

Order *Ornithischia*

Group *Ceratopsian*

LENGTH *2 m (6.5 ft)*

WEIGHT *50 kg (110 lb)*

NAME MEANING *"Parrot Lizard"*

FOUND *China, Mongolia, Russia*

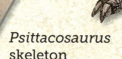

Psittacosaurus skeleton

FOOTPRINT IDENTIFICATION

FASCINATING FACT

A 13 cm (5 in) baby *Psittacosaurus* has been found inside the fossilized stomach of a prehistoric badger-like mammal called a *Repenomamus*, proving that small dinosaurs made tasty meals for ancient mammals.

FOSSIL STUDY

Psittacosaurus skull

When the first *Psittacosaurus* specimens were discovered in Mongolia in 1922, one of the first things that scientists noted was their square skulls and curved beaks. The protruding nasal bone, called a rostral bone, is only found in Ceratopsians! The hooked beak, similar to a parrot's, gave *Psittacosaurus* its name, which means "parrot lizard."

Psittacosaurus is a very important Ceratopsian. Paleontologists have discovered more than 400 specimens spanning all ages, including some that are complete! All of these fossils allow scientists to piece together a full picture about the habits, lifestyle, and life cycle of a typical *Psittacosaurus*. This dinosaur has more species of dinosaurs in its group than any other dinosaur; scientists have so far identified up to eleven different species of *Psittacosaurus* across Siberia, Asia, and Southeast Asia.

Psittacosaurus is unique in the Ceratopsian family because it was bipedal and quite small at only about 0.6 m (2 ft) tall and weighing about as much as a medium-sized dog. It had short, grasping arms and longer, muscular hind legs, as opposed to the pillar-like legs of *Triceratops* or *Styracosaurus*. Another noticeable difference between *Psittacosaurus* and other Ceratopsians is that *Psittacosaurus* had no horns or fancy frill. You may be wondering how a group of "horned dinosaurs" could include a dinosaur that had no horn! The one trait that ALL Ceratopsians have is actually not their horns but their unique parrot-like beaks.

Psittacosaurus also had bristles running along its tail. They were grouped together in bundles and were stiffer and stronger than bird feathers, similar to the "beard" of a turkey. These quill-like bristles would not have been much help in defense and were only used for display, but what a magnificent display it would have been!

STYRACOSAURUS

[sty–RAK–uh–SOR–us]

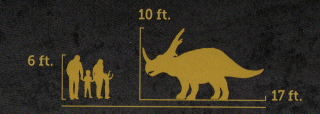

10 ft.

6 ft.

17 ft.

CLASSIFICATION

Order _Ornithischia_

Group _Ceratopsian_

LENGTH _5.2 m (17 ft)_

WEIGHT _2,450 kg (5,400 lb)_

NAME MEANING _"Spiked Lizard"_

FOUND _Alberta, Canada_

FOOTPRINT IDENTIFICATION

Styracosaurus was a moderately-sized quadruped that lived in the floodplains and woodlands of western North America. This low-browsing herbivore feasted on the ferns and cycads of its habitat, using its stout, strong body and horns to knock down small trees to reach thicker brush and undergrowth. Its parrot-like beak could grasp and tear vegetation, while its teeth were ideal for grinding up plants. When the teeth became worn down and fell out, new ones continually replaced them!

Styracosaurus, whose name means "spiked lizard," not only had an impressive neck frill, but it also had a long horn atop its nose and multiple horns extending out from its cheeks and neck frill. These structures gave it an intimidating appearance that would have kept many potential predators from attempting an attack.

FASCINATING FACT

In classical Greece, warriors carried spears that had a steel spike on the end called a styrax. *Styracosaurus* got its name from this deadly spear topper.

Most of the *Styracosaurus* fossils that have been discovered were found in bone beds in Alberta, Canada. Bone beds were formed when numerous specimens were washed into the areas by flood waters, or when the dinosaurs living in the areas all died at the same time from some cataclysmic event (like a drought or flood). In 1915, a nearly complete specimen of *Styracosaurus* was found, but in the excitement, the location of the bone bed was lost for nearly 100 years! It wasn't until 2006 that the site was rediscovered, complete with skull fragments that had been left behind by the previous dinosaur hunters.

Styracosaurus horns (right)

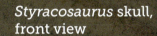

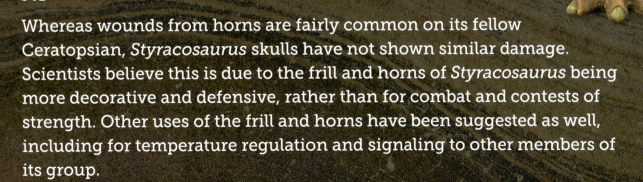

Styracosaurus skull, front view

FOSSIL STUDY

Whereas wounds from horns are fairly common on its fellow Ceratopsian, *Styracosaurus* skulls have not shown similar damage. Scientists believe this is due to the frill and horns of *Styracosaurus* being more decorative and defensive, rather than for combat and contests of strength. Other uses of the frill and horns have been suggested as well, including for temperature regulation and signaling to other members of its group.

TRICERATOPS

[tri–SAIR–uh–tops]

The great *Triceratops*, one of the most recognizable dinosaurs in the world, earned its name, which means "three-horned face." *Triceratops* was a plant eater, but that doesn't mean it was a gentle, peaceful dinosaur! It was a massive beast, measuring up to 9 m (30 ft) in length, and was as tall as a one-story building! From its muscular, trunk-like legs to its three menacing horns and thick, stout body, this dinosaur is considered one of the most powerful dinosaurs of its era.

Triceratops is considered one of the most common dinosaurs, its fossils representing many different species. Common to all species was the short, bony frill that rose up from the neck. This frill probably had many uses. It could have been used as a signal to other *Triceratops*, possibly even flushing pink due to blood vessels under the skin. It also could have been used as a defensive shield against predators or even against other *Triceratops* as the males locked horns in competitions of strength.

Fossilized *Triceratops* horn

CLASSIFICATION

Order *Ornithischia*

Group *Ceratopsian*

LENGTH *9.1 m (30 ft)*

WEIGHT *10,000 kg (22,000 lb)*

NAME MEANING *"Three-horned Face"*

FOUND *Colorado; Montana; Wyoming; South Dakota; Alberta, Canada; Saskatchewan, Canada*

Roaming the marshes and forests of western North America during the same time period as *Tyrannosaurus rex*, *Triceratops* is thought to have been a favorite meal of *T. Rex*. But it would not have been an easy meal, for *Triceratops* could fight off the best of the hunters with its formidable horns. *Tyrannosaurus rex* would have had to attack from the side or behind to avoid *Triceratops*' swordlike horns and to bypass the defensive protection of its large neck frill. Teeth marks from *Tyrannosaurus rex* on some *Triceratops* bones, however, reveal that *Triceratops* lost many of these battles.

FOSSIL STUDY

In Montana in 2006, a *Triceratops* and a juvenile *Tyrannosaurus rex* were found encased in sediment where they had died, possibly while in battle. This find has been given the name "Dueling Dinosaurs." What is remarkable is that both sets of bones are some of the most complete specimens of these dinosaurs ever found. The *Triceratops* fossils even preserved skin impressions on its hips and frill!

Triceratops skull

Triceratops foot bones

FASCINATING FACT

Triceratops is the official "State Fossil" of South Dakota and the official "State Dinosaur" of Wyoming!

CERATOPSIANS GROUP GUIDE

CHARACTERISTICS

Ceratopsians are distinguished from other dinosaurs by their horn-covered, parrot-like beaks. The most famous members of this "horned-face" group had a distinctive neck frill and elaborate horns on their faces.

FOSSIL SITES

Ceratopsian dinosaur fossils have been found in North America, Europe, and Asia.

SIZE COMPARISON

12 ft.

6 ft.

Psittacosaurus

Protoceratops

Styracosaurus

Triceratops

ORNITHOPODS

CORYTHOSAURUS

[ko–RITH–oh–SOR–us]

6 ft. 11 ft. 30 ft.

Corythosaurus is an Ornithopod dinosaur, specifically part of a sub-group called hadrosaurs, or duck-billed dinosaurs. These large herbivores all display a hollow crest or helmet-like structure on their heads, likely used for ornamentation and communication. The unique crest of *Corythosaurus* looks very similar to the helmets worn by ancient Greek soldiers from Corinth, thereby giving *Corythosaurus* its name.

Corythosaurus lived in a woodland environment, possibly near swampy areas. Because it had a fragile, short beak, early speculation was that it lived in or near water and fed mainly on soft vegetation. New evidence, including specimens found with pine needles, seeds, twigs, and fruit, reveal that *Corythosaurus* would have been competing with other low-grazing herbivores for the ground-level vegetation. In dry times, though, *Corythosaurus* could have moved to the swamps to find water vegetation.

Corythosaurus, like all other dinosaurs, had bony rings around its eyes known as scleral [SKLAIR–uhl] rings. The scleral rings of *Corythosaurus* are very similar to those of modern birds and reptiles, evidence that *Corythosaurus* would have been active only for short periods, but all throughout the day and night.

FOSSIL STUDY

Fossilized *Corythosaurus* skin imprint

Scientists believe that the hollow crest of *Corythosaurus* may have been a complex tool for communicating with other members of its species. When air passed through these structures, which were connected to the nasal passages, sound reverberated and was amplified, much like a horn. These loud calls could have been warnings or used in mating.

CLASSIFICATION

Order *Ornithischia*

Group *Ornithopod*

LENGTH *9 m (30 ft)*

WEIGHT *4,500 kg (10,000 lb)*

NAME MEANING *"Helmet Lizard"*

FOUND *Alberta, Canada; Montana; Colorado; Utah*

FOOTPRINT IDENTIFICATION

Corythosaurus skull

FASCINATING FACT

A collection of *Corythosaurus* fossils—two nearly complete specimens—were being shipped across the Atlantic on a merchant ship in 1916 during World War I. The ship was sunk by a German cruiser, and all cargo, including the precious *Corythosaurus* remains, was lost.

HETERODONTOSAURUS

[HET-er-oh-DONT-oh-SOR-us]

6 ft. 3 ft.

5.75 ft.

Not all dinosaurs were gigantic, lumbering Sauropods or huge, fearsome Theropods. Some dinosaurs were as small as turkeys when full grown! *Heterodontosaurus* was an early Ornithischian (bird-hipped) dinosaur. Ornithischians were typically plant eaters, but there is some debate about whether this quick, little African dinosaur was an herbivore, a carnivore, or possibly even an omnivore.

The discussion about the eating habits of *Heterodontosaurus* comes from the variety of teeth that it had. Most reptiles and dinosaurs have only one kind of tooth, but *Heterodontosaurus* had three types! Each type of tooth had a separate job. The numerous teeth lining both sides of the back of the jaws were molar-like and were used for grinding, similar to your own back teeth. In the front of the jaws were a few small, pin-shaped teeth used for snipping and slicing vegetation. The most distinctive teeth were two pairs of long canine teeth called tusks that filled the gap between the front and back teeth.

Heterodontosaurus teeth

FOSSIL STUDY

Scientists are still not sure what *Heterodontosaurus* used its tusks for. One theory is that they were actually used as a tool for digging. Perhaps *Heterodontosaurus* needed to dig up underground roots or tubers or to dig into termite mounds for a tasty feast! Another theory is that they were somehow used in defense, warding off other males.

Heterodontosaurus skull

CLASSIFICATION

Order   *Ornithischia*

Group   *Ornithopod*

LENGTH   *1.2–1.75 m (4–5.75 ft)*

WEIGHT   *3.5 kg (7.5 lb)*

NAME MEANING   *"Different Teeth Lizard"*

FOUND   *Southern Africa*

FOOTPRINT IDENTIFICATION

Dinosaurs related to *Heterodontosaurus* had long, coarse bristles similar to a mammal's thick fur, which has caused some scientists to speculate that this Ornithopod did as well. However, there is nothing from the recovered specimens of *Heterodontosaurus* to suggest the presence of bristles.

FASCINATING FACT

The complete skull of a *Heterodontosaurus*, one of the smallest dinosaur skulls in the world, was discovered in a drawer in the Iziko South African Museum in 2008. It had been stored away there, unidentified and forgotten, since the 1960s!

IGUANODON

[ih–GWAN–oh–don]

13 ft.
6 ft.
33 ft.

CLASSIFICATION

Order *Ornithischia*

Group *Ornithopod*

LENGTH 10 m (33 ft)

WEIGHT 5,000 kg (11,000 lb)

NAME MEANING *"Iguana Tooth"*

FOUND *Europe (Belgium, Germany, England)*

FOOTPRINT IDENTIFICATION

The first discovered *Iguanodon* fossil was a tooth found in a quarry in England in 1822 by Dr. Gideon and Mary Ann Mantell. Because *Iguanodon* was only the second dinosaur to have been discovered and named, there was not a lot known about the massive beast. At first, scientists believed it to be a fish, then a rhinoceros, then a huge reptile, before realizing it was an entirely unknown creature—a dinosaur. Because its tooth looked so much like that of an iguana's, Dr. Mantell named the animal *Iguanodon*, "iguana tooth."

Iguanodon is an Ornithopod, an herbivore characterized by walking on two legs, having no armor, and having three-toed feet. Even so, it is now believed that *Iguanodon* mostly walked on all four legs, although the animal certainly could have raised up on its back two legs for walking or grazing on high branches. This ability would have allowed it to see and smell farther than a quadruped. *Iguanodon* had a long head, like a horse, and its muscular structure supports the idea that it had an extremely long tongue. Its tail was inflexible, stiffened by bony tendons.

Iguanodon was a herd animal, traveling in groups that made use of their large numbers and speed to evade enemies. *Iguanodon* often grew very large, larger than its carnivorous predators, which gave it another layer of defense.

Iguanodon foot

FASCINATING FACT •••

Iguanodon's thumb spike was initially believed to be a horn. When the animal was first reconstructed, the spike was placed on its nose!

Iguanodon hand

FOSSIL STUDY

Iguanodon's curious spiked thumb could be wrapped around branches and used to bring them down to its mouth to eat. The thumb also could have been used to break open nuts or seeds, or as dagger-like protection against predators.

MAIASAURA

[MAY–uh–SOR–uh]

A latecomer to the paleontology world, the first *Maiasaura* was not discovered until 1978. The dinosaur hunters that found it gave it a name that means "good mother lizard." This is an appropriate name because they discovered it in a dinosaur nesting site, which is an area where dinosaurs grouped together to lay their eggs in nests. Many fossilized eggs, embryos, and young *Maiasaura* were found together in this site in Montana and helped prove that some dinosaurs were social animals.

CLASSIFICATION

Order *Ornithischia*

Group *Ornithopod*

LENGTH *9 m (30 ft)*

WEIGHT *3,600 kg (8,000 lb)*

NAME MEANING *"Good Mother Lizard"*

FOUND *Montana*

FOOTPRINT IDENTIFICATION

6 ft.

9 ft.

30 ft.

FOSSIL STUDY

Hundreds of fossils at all stages of life have been discovered at the large nesting site in Montana. *Maiasaura* nests were made in hollowed-out ground about 7 m (23 ft) apart with soft vegetation lining the nest. Each nest would contain as many as 30–40 eggs laid in a circular pattern, each egg being about the size of a grapefruit. Because the mother *Maisaura* was too big to sit on the nest, the eggs were kept warm by the rotting vegetation in the nest.

Maiasaura skull

Diorama of *Maiasaura* nest

Maiasaura was a large Ornithopod, specifically a part of the family of hadrosaurs, or "duck-billed dinosaurs," so-called because they have beaks like ducks! These big plant eaters not only made their nests in groups but also lived in large herds. Each herd may have had as many as 10,000 *Maiasaura* dinosaurs! An adult *Maiasaura* needed up to 90 kg (200 lbs) of food each day, so these enormous herds would have needed to stay on the move to find enough to eat.

A newly hatched *Maisaura* was quite helpless. Its bones and joints were not fully formed, and it took some time for its body to be ready to leave the nest. Meanwhile, its mother would need to bring it food and protect it. Scientists think it may have taken several years before a *Maiasaura* was ready to be completely independent.

FASCINATING FACT

The site in Montana where the nesting site and the *Maiasaura* specimens were found is now called Egg Mountain in honor of the "good mother lizard."

ORNITHOPODS GROUP GUIDE

CHARACTERISTICS

The group of dinosaurs known as Ornithopods, also known as the "bird-feet" dinosaurs, walked upright and on their toes like birds. Ornithopods had horny beaks that cropped off plants. They often lived in herds.

FOSSIL SITES

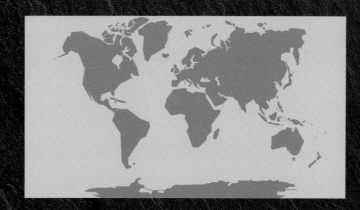

Fossils of Ornithopod dinosaurs have been found on every continent.

SIZE COMPARISON

13 ft.

6 ft.

Heterodontosaurus

Maiasaura

Corythosaurus

Iguanodon

PACHYCEPHALOSAURS

...ycephalo... ...dinosaurs with...
...ful... ...thick, do...
...er dinos...
...osaurs belonging t...
...Pachycephalosaurus... ...includ...
...and Stegoceras. Pach... ...dal lived i...
...nosaurs with power... ...d legs and...al...
...s. Their thick, dom...ke structure... n th...
...hich was nor u...ommon in P...
...d them the...
...d to head...ut other dinosaurs, e...em the...
..."boneheads." Pachycephalosaurs...were quick
...Pachycephalosaur foss...s hav...been fou...
...oughout North Amer...a...

PACHYCEPHALOSAURUS

[PACK-ih-SEF-uh-luh-SOR-us]

6 ft.   7 ft.

18 ft.

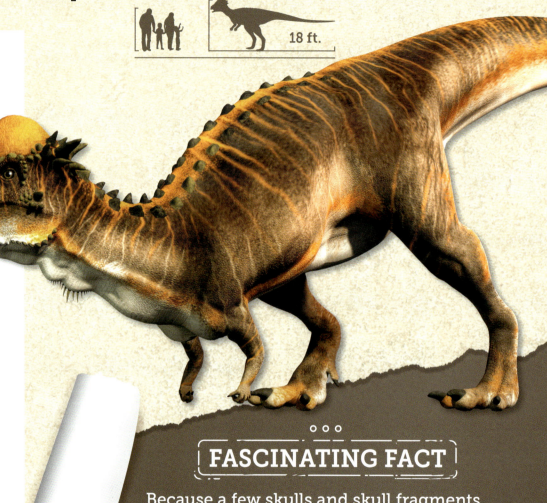

CLASSIFICATION

Order _Ornithischia_

Group _Pachycephalosaur_

LENGTH _5.5 m (18 ft)_

WEIGHT _1,800 kg (4,000 lb)_

NAME MEANING _"Thick-Headed Lizard"_

FOUND _Montana; South Dakota; Wyoming; Saskatchewan, Canada; Mongolia_

FOOTPRINT IDENTIFICATION

FASCINATING FACT

Because a few skulls and skull fragments were the only _Pachycephalosaurus_ bones found for many years, paleontologists initially thought that their unique skulls were actually dinosaur kneecaps!

FOSSIL STUDY

In addition to having a nearly 23 cm (9 in) thick dome on the top of its skull, *Pachycephalosaurus* had other unique cranial features. Bony knobs encircled the back and sides of the skull, while blunt spikes protruded out from around its snout. Scientists speculate that juveniles may have had flatter skulls that thickened and rounded out on top as they matured.

Pachycephalosaurus skull

*P*achycephalosaurus was a bipedal herbivore that belonged to a group of dinosaurs called Pachycephalosaurs, nicknamed the "bone-headed dinosaurs." They got this name due to their thick skulls made of bone, some of which were around 25 cm (10 in) thick! *Pachycephalosaurus* is considered the largest of the boneheads, with five-fingered claws on short arms, long hind legs, and a stiff tail that provided balance.

Pachycephalosaurus is a dinosaur whose popularity outshines the scant number of discovered fossils and how much scientists actually know about it! Very few fossils of *Pachycephalosaurus* have been found, and those are mostly skulls and skull fragments. The thick bony structure of the skulls has kept them much better preserved than the rest of the fragile skeleton. Little, therefore, is known about what the body of *Pachycephalosaurus* looked like, so scientists have filled in the missing information with similar dinosaurs.

Pachycephalosaurus is a favorite with many people because it did something that no other dinosaur family did—it used its bony head to headbutt other members of its group or predators! In the past, scientists have debated whether *Pachycephalosaurus* actually engaged in this behavior, but recent studies have shown that damage consistent with headbutting is present on many of the existing skulls.

STEGOCERAS

[STEG-oh-SAIR-us]

Top of *Stegoceras* dome

6 ft.

3 ft.

6.6 ft.

CLASSIFICATION

Order *Ornithischia*

Group *Pachycephalosaur*

LENGTH 2 m (6.6 ft)

WEIGHT 10–40 kg (22–88 lb)

NAME MEANING "Horned Roof"

FOUND Alberta, Canada; Saskatchewan, Canada; New Mexico; Montana

FOOTPRINT IDENTIFICATION

Stegoceras was a bipedal herbivore in the Pachycephalosaur family. It was smaller than some other bone-headed dinosaurs, only about the size of a goat. It might be easy to get the name of *Stegoceras* confused with *Stegosaurus*, but the lightweight, bipedal dinosaur with the domed head looks very different from the stout quadruped with armor! *Stegoceras* lived in the forests of North America, specifically from New Mexico all the way north into Alberta and Saskatchewan, Canada.

Similar to *Pachycephalosaurus*, for many decades the only *Stegoceras* fossils that had been discovered were fragments from skulls. That changed in 1902 when a nearly complete skeleton was found. This was the first skeleton of *Stegoceras* that was more than just a skull or a few random bones. This makes *Stegoceras* one of the best-known dinosaurs and still one of the few Pachycephalosaurs of which we have any fossils other than the skull.

The distinctive domed skull of *Stegoceras* was about 10 times the thickness of a human skull. It was triangular in shape when looking from the side, with a ridge around the back of the skull. Adult *Stegoceras* skulls have two kinds of domes. One kind is thicker and heavy, which paleontologists believe belonged to males, and another kind is flatter and thinner, which are believed to be female skulls.

Underside of *Stegoceras* dome

FOSSIL STUDY

Scientists have discovered four alternating layers of bone in the skulls of *Stegoceras*—three layers that were stiff and solid and one layer that was soft and spongy! This construction was a perfect setup for headbutting, with the hard outer layer acting like a hard hat to protect the brain and with the inner spongy layer absorbing the impact. Scans of *Stegoceras* skulls have shown damage consistent with headbutting.

Stegoceras skull

FASCINATING FACT ∘∘∘

Even though *Stegoceras* is considered an herbivore, based on the size and type of teeth it had, paleontologists have not ruled out the possibility that the quick-footed dinosaur may have eaten insects and other small creatures.

PACHYCEPHALOSAURS GROUP GUIDE

CHARACTERISTICS

Pachycephalosaurs were quick, bipedal dinosaurs with powerful hind legs and small forearms. The thick, dome-like structure on their heads, which was used to headbutt other dinosaurs, earned them the nickname "boneheads."

FOSSIL SITES

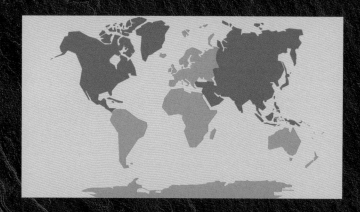

Fossils of Pachycephalosaurs have been found in North America and Asia.

SIZE COMPARISON

6 ft.

7 ft.

Stegoceras

Pachycephalosaurus

STEGOSAURS

KENTROSAURUS [KEN–troh–SOR–us]

CLASSIFICATION

Order _Ornithischia_

Group _Stegosaur_

LENGTH 5 m (16 ft)

WEIGHT 1,100 kg (2,425 lb)

NAME MEANING _"Spiked Lizard"_

FOUND Tanzania, Africa

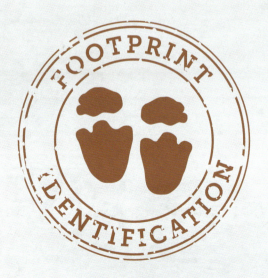

FOOTPRINT IDENTIFICATION

Tibia (lower leg bone) of a _Kentrosaurus_

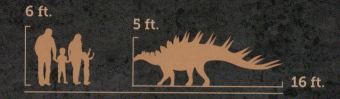

6 ft.   5 ft.   16 ft.

The _Kentrosaurus_ had a bill-shaped head and large snout. It had large nasal passages and a large olfactory bulb, which is the structure located just above the nasal cavity that sends signals to the brain about smells. These two features would have given _Kentrosaurus_ an excellent sense of smell. As an herbivore it didn't need large, strong teeth. Instead, it had many tiny teeth tucked into its cheeks.

FASCINATING FACT •••

The skull and spine of the only remaining complete _Kentrosaurus_ model were thought to have been destroyed or lost during World War II. The missing bones were eventually found in a basement closet drawer!

FOSSIL STUDY

Spikes abound on this Stegosaur! The longest spikes were found running down the entire length of its tail, which it could whip around to defend itself against an attacking enemy. This herding animal could, therefore, band together with other *Kentrosauruses* to create a protective, deadly wall with their spikes.

Illustration of *Kentrosaurus* tail spikes

This quadruped had double rows of broad plates from its neck down to its mid-back, and from there the rows continued with long spikes (from which it gets its name) down to the very end of the tail. The tail itself was flexible, not having the bony tendons sometimes seen in other types, and may have been swung around, using the spikes to protect against predators. The *Kentrosaurus* has two spikes that are approximately double the size of its back spines. Some scientists believe that they attached onto the shoulders to help protect its front while others believe they were on either hip in case a predator tried to get around its tail.

Many bones from this curious Stegosaur were discovered in a dig in the early 1900s. Located in Tanzania, not far from the coast of the Indian Ocean, the Tendaguru Formation has provided thousands of *Kentrosaurus* bones. There have been no complete skeletons found, but the plethora of bones found has allowed scientists to reconstruct two full *Kentrosaurus* specimens. Unfortunately, one of the specimens was destroyed by Allied bombing during World War II.

STEGOSAURUS

[STEG-uh-SOR-us]

6 ft. 11 ft. 29.5 ft.

Stegosaurus is the largest and most well-known dinosaur in the Stegosaur family, which includes other plated and spiked dinosaurs. *Stegosaurus* is very recognizable to dinosaur fans with its double row of triangular-shaped plates running along its spine and its spiky tail. It was a heavy, round-backed herbivore with short legs, a beak, and small, rounded teeth. Because of its short legs and weak jaws, *Stegosaurus* fed on low, soft vegetation such as ferns, mosses, and fruits.

The spine plates of *Stegosaurus* are not thought to have been used for defense since they were too fragile to provide much protection. It is believed the plates were used either for appearance—a colorful display allowing *Stegosaurus* to recognize others of its kind—or to help the animal regulate its body temperature. The plates were not attached to the spine itself but embedded in the skin, likely arranged in a double row with the plates standing upright.

CLASSIFICATION

Order *Ornithischia*

Group *Stegosaur*

LENGTH *9 m (29.5 ft)*

WEIGHT *5,200–7,000 kg (11,500–15,400 lb)*

NAME MEANING *"Roof Lizard"*

FOUND *North America (Wyoming, Utah, Colorado)*

FOOTPRINT IDENTIFICATION

Stegosaurus skeleton

As new discoveries provide more information about a dinosaur's anatomy, lifestyle, and habits, paleontologists must continually modify their conclusions. For example, some early paleontologists thought that *Stegosaurus* had two brains—one in the skull and one in the hip! However, scientists now believe that the space near *Stegosaurus*' hip was used for the storage of food during times in which food was scarce. Another theory that has been altered is the early belief that *Stegosaurus* walked on two legs. Scientists now conclude that the animal's short legs and heavily built body make this very unlikely.

FOSSIL STUDY

Thagomizer (tail spikes) of a *Stegosaurus*

The tail spikes of *Stegosaurus*—having the intimidating name thagomizer—were a deadly defensive weapon against enemies. Each of the four spikes, which extended out horizontally from the tail, was 60–90 cm (2–3 ft) long! Many of the fossilized spikes found are damaged at the tip, evidence of violent combat. Fossils from dinosaurs such as the *Allosaurus*, a predator to the *Stegosaurus*, have been found with fatal injury marks from a *Stegosaurus*' thagomizer.

FASCINATING FACT

In 1982, after a two-year effort, Colorado fourth-graders successfully convinced the governor of Colorado to officially declare the *Stegosaurus* the state dinosaur of Colorado!

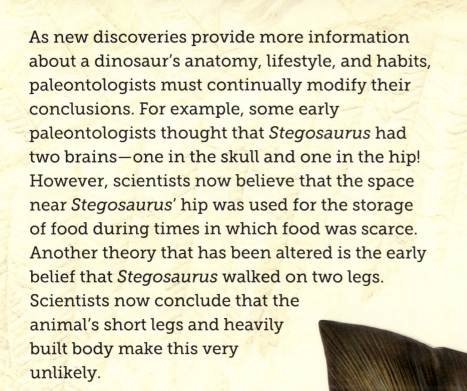

TUOJIANGOSAURUS

[too–YANG–oh–SOR–us]

6 ft. 7 ft. 23 ft.

When a nearly complete *Tuojiangosaurus* skeleton was discovered in 1974 during the construction of a dam in China's Sichuan Valley, the specimen became the first Stegosaur ever to be discovered in China. The Sichuan Valley had been a vast floodplain with many rivers and lakes. *Tuojiangosaurus*, an herbivore with spoon-shaped teeth in the front of its mouth, was well suited for the soft, low-lying vegetation of this habitat.

Like other Stegosaurs, *Tuojiangosaurus* had vertical spines running along its back, spikes at the end of its tail, a long, narrow head, and a stocky body with a rounded back. Its pillar-like legs were short, with its hind legs longer than its front legs. This meant that the animal held its head rather close to the ground. It had unique shoulder spikes, the length of which may have demonstrated *Tuojiangosaurus'* health and vitality when seen by other dinosaurs.

FASCINATING FACT

Tuojiangosaurus had a flexible tail that could be whipped around quickly. Like other Stegosaurs, it used its thagomizer in defense!

FOSSIL STUDY

Tuojiangosaurus had a long, narrow, flat head. The tip of its snout was actually a horn-covered beak, similar to that of a turtle! Its beak would have been very useful in snipping off small bits of choice vegetation while avoiding any it did not want.

Tuojiangosaurus skull

CLASSIFICATION

Order *Ornithischia*

Group *Stegosaur*

LENGTH *7 m (23 ft)*

WEIGHT *2,800 kg (6,200 lb)*

NAME MEANING *"Tuo River Lizard"*

FOUND *China*

Tuojiangosaurus spine plates

FOOTPRINT IDENTIFICATION

Similar to other armored dinosaurs in the Stegosaur family, *Tuojiangosaurus* had triangular-shaped plates running along its spine. Unique to *Tuojiangosaurus*, though, were the smaller size and sharper edges of these plates. The plates positioned directly over the hips were the largest and most pointed, the others decreasing in size and becoming more round in shape as they neared the animal's head. Even though these plates looked like a fierce defensive weapon, they would not have provided much protection at all, as they were too brittle. They were likely only for display, as a way to signal to other members of *Tuojiangosaurus*' family.

STEGOSAURS GROUP GUIDE

CHARACTERISTICS

Stegosaurs were quadrupedal herbivores with short legs that kept them close to the ground. The bodies of Stegosaurs were protected by double rows of bony plates and spikes. They had small heads with hard beaks, useful for chomping plants.

FOSSIL SITES

Stegosaur fossils have been found in North America, South America, Africa, Europe, and Asia.

SIZE COMPARISON

6 ft.

11 ft.

Kentrosaurus

Tuojiangosaurus

Stegosaurus